István Csuzi

new series: Studies about Public Transportation Volume 2

István Csuzi

new series: Studies about Public Transportation Volume 2

Bus systems efficiency indicators

LAP LAMBERT Academic Publishing

Cover image: www.ingimage.com

Publisher:
LAP LAMBERT Academic Publishing
is a trademark of
International Book Market Service Ltd., member of OmniScriptum Publishing Group
17 Meldrum Street, Beau Bassin 71504, Mauritius

Printed at: see last page
ISBN: 978-613-7-32919-1

FOREWORD,

replacing the introduction…

Lot of studies was done by updating the data and interpretations of the author's Ph.D. Thesis: *"Contributions to the evaluation and optimization of the energy performance and availability of the urban electric traction system"*, coordinated by Prof. Dr. Eng. Ioan FELEA at the University of Oradea sustained in 2011. Even the words on the introductory page of Ph.D. work suggest that this concern will continue and will generate new opportunities:

"There are moments in life when you stop and look back: where did you go, on what road you came. Public support for this doctoral thesis may be such a moment. Maybe I can see the continuity of my interest in science from in high school and college, respectively in the work places, there were moments of real joy, but we also counted less "resounded" successes. The road was not easy, the efforts made did not always materialize thanks to perseverance and consistency have sustained my efforts, but on my own I have not been able to continue without a good team, without real help from my family, colleagues and mentors. Maybe alone not been able to finish this stage but Prof. Dr. Eng. Ioan FELEA was the decisive factor in the birth of this work, without his explanations, advice and co-ordination perhaps I fail the job. I can not to remember the big help received from Prof. Dr. Eng. József LINGVAY. After nearly twenty years of absence at the instigation of him, I have again gained the experience and satisfaction of attending the scientific conferences. During the elaboration of the works, the measurements and the verifications made were possible through an efficient organization, especially through the contribution of colleagues from Oradea Transport Local SA (Liviu TUDUC, Ioan MUREŞAN, Sebastian SILAGHI, László LIBOR and many others ....).

If I could stop here, I would do ... but life goes on! The list to which the contribution the help, the understanding of the present work is incomplete, and I know it remains an open list for the future. I hope and I am convinced that the work

1

does not end here, this thesis is just the completion of the first stage in an ongoing process, ..., there are thousands of ways and possibilities for research to improve and optimize the energy efficiency of urban public transport, new areas of application of results will be found through direct implementation in production, exploitation for the immediate benefit of current and future research. (István CSUZI, Oradea, 2011)"

There have been many changes in previous years, achievements have been recorded in the public transport company in Oradea, but we also regret the fact that not all Romanian municipalities have retained their tram networks, Brasov, Constanta, Sibiu and Resita in the last ten they took the decision to remove Trams from urban traffic. There are 11 county residences, where the tradition continues, some of them even having a real chance of developing the systems, among which Oradea is recognized as a city with one of the well good maintained Tram system.

As a personal remark, after more than 8 years as General Director of Oradea Transport Local SA and Associate Professor of University of Oradea studies are cumulated, a lot of works was presented in difference Conferences in 6 countries. All of them has a common scope and goal to optimize the public transport systems, to give new methods and technologies for daily work of the sector. Thanks to this effort I had the chance to promote the ideas, as a "Light Rail Ambassador" in my role of President of Light Rail Assembly and Vice President of UITP- International Public Transport Association based in Brussels. Start from May 2017 my luck given me the chance to try my skills and abilities in Middle East as General Manager of City Transport, the public transport company of Abu Dhabi. It's a big challenge how to optimize energy efficiency in different weather and social conditions, the specific research activity will continue related to those factors.

Maintaining the quality of light rail transport in the city of Oradea, even the development of the network is due to the symbiosis in which travelers use trams, of the total daily passenger traffic of approx. 150,000 people over 60% use the tram. Why choose the Tram and not the Electric Bus or Trolleybus? Why do not use a liquid fuel combustion bus? There are questions that are difficult to answer easier in two words, such as being greener or having greater efficiency. The simplicity of the answer is articulated by the complexity of systems, a multi-criteria analysis of economic, technical and social aspects will allow us to have better idea of the reality of urban public transport with electric traction on the ground, why is high and major impact on life the public transportation.

This work is mainly scientific, intended for specialists in the field, students from technical faculties with specializations in energy, electrical engineering, but can be useful through public transport systems presentations starting from feasibility study makers to the decision-makers from public administrations.

The book is the second volume of the new series of „STUDIES ABOUT PUBLIC TRANSPORTATION". There is a permanent debate between prfessionals about Trams and Buses, which have priority on the focus of developments of public transport. Main issue are related to the efficiency of recommended systems are relating to transport capacity adaptation to real needs, availability of fleet, energy savings were scopes of many studies. "BUS SYSTEMS EFFICIENCY INDICATORS" is a short approach of key performance indicators (KPI) for basic characteristics. Chapters are integral presentations of the scientific papers on this topic, presentations at symposiums and conferences both in the country and abroad are listed in the list of publications in the appendix at the end of the book.

Abu Dhabi,   István CSUZI, Ph.D.

20th December, 2017

ABSTRACTS:

Bus systems efficiency indicators

STUDY 1 - Method for energy efficiency analysis of a public transport system by the case study of Oradea Transport Local SA Bus and Tram network analysis

This study propose an analysis method of energy efficiency comparisons of the urban public transport system through the case study applied in Oradea. The study contains a brief description of Public Transport Company and network in Oradea, passenger statistics, analysis of energy consumption, passenger/kilometer parameter evaluation, compare liquid fuel and electricity consumption, recommendations on optimizing energy consumption. The comparisons are focusing on OTL's own vehicle fleet (trams and buses) and the energy performance of the system.

STUDY 2 - The urban electric bus, a sustainable solution to increase energy efficiency of public transport and reduce atmospheric pollution in the cities

Analysis of the values of public transport energy indicators and metropolitan air pollution together with the branch specific economic forecasts, the authorities and suppliers can estimate future trends. To reduce the urban usage of personal automobiles it is needed to increase the proportion of electric public transport vehicles (Tramways, Trolleys, Electric buses), also it needs to increase the urban mobility satisfaction by increasing the number of trips by public transport, all together can insure us more livable cities.

Study 1: **Method for energy efficiency analysis of a public transport system by the case study of Oradea Transport Local SA Bus and Tram network analysis**

1.1.   General notions about public transport systems efficiency analyzis

The Public Transport Company in Oradea (Oradea Transport Local SA) currently operates seven tram and twenty urban bus lines, covering the city's main transport axes [2]. In addition to the local public transport organization and exploitation OTL SA provides the public transport services for two metropolitan-area villages (Sânmartin and Borş) and administratively associated villages of them on six bus lines, and operates an international cross-border bus route (between Oradea/Romania and Biharkeresztes/Hungary). The study aims to compare the power consumption of trams and buses in perspective of distances and the transported passengers increasing energy efficiency and attractiveness of public transport for passengers. The increase of energy efficiency and increase of transported passenger clearly leads to an improvement of economic indicators.

The analyzed period regarding to the energy efficiency are 2014, 2015 and 2016. Yearly, monthly, weekly and daily analyzes. In order to analyze the passenger flow evaluation we are monitoring the number of passengers in each hour of days. Based on this analyzes we can determine and optimize the required vehicle capacity.

The bus and tram mileage, covered distances are gained from two applications. These applications are: "Fleet Management" software (which generates data based on GPS tracking) and an "in-house" developed software that is specifically designed to evaluate the company's bus fleet (mileage, consumption data and the calculation of worked hours).

Passenger flow data is extracted from two applications. The "E-ticketing" system for RFID-Card validation, as well as the "GPS based Fleet Management" application for paper based ticket validation. Based on the comparison and analyze of these data we can estimate the required vehicle capacity for different periods of the day.

The electric energy consumption data (weekly, monthly and yearly) are available separately for each electric rectifier (power) station. The diesel fuel consumption data for the analyzed period is available from the "in-house" bus fleet analyzing software. The software is using a consumption database with exact measurement for all the bus-lines and all type of busses of the fleet. The Diesel fuel consumption is calculated based on these data.

To compare our liquid fuel and electric power consumption we need a common denominator [3].

Table 1.1.: *Energy value standard comparison in TOE*

Electric Energy	1 MWh = 0.086
Thermal Energy	1 G-Cal = 0.1
Natural Gas	1000 Nm³ = 0.805
Black Oil	1 to = 0.95
Filtered Oil	1 to = 0.97
Fuel	1 to = 1.05
Diesel	1 to = 1.015

The **T**one of **O**il **E**quivalent (TOE) is a unit of energy defined as the amount of energy released by burning one tone of crude oil. It is approximately 42 gigajoules, although as different crude oils have different calorific values, the exact value is defined by convention; several slightly different definitions exist. The International Energy Agency defines one tone of oil equivalent (TOE) to be equal to 41.868 GJ (11.63 Megawatt hours). Different energy value conversions were done based on the International Energy Agency calculator. [4]

1.2. The urban public transport system in Oradea city and neighbor regions

Urban bus and tram network map show a very good coverage of Public Transport accessibility.

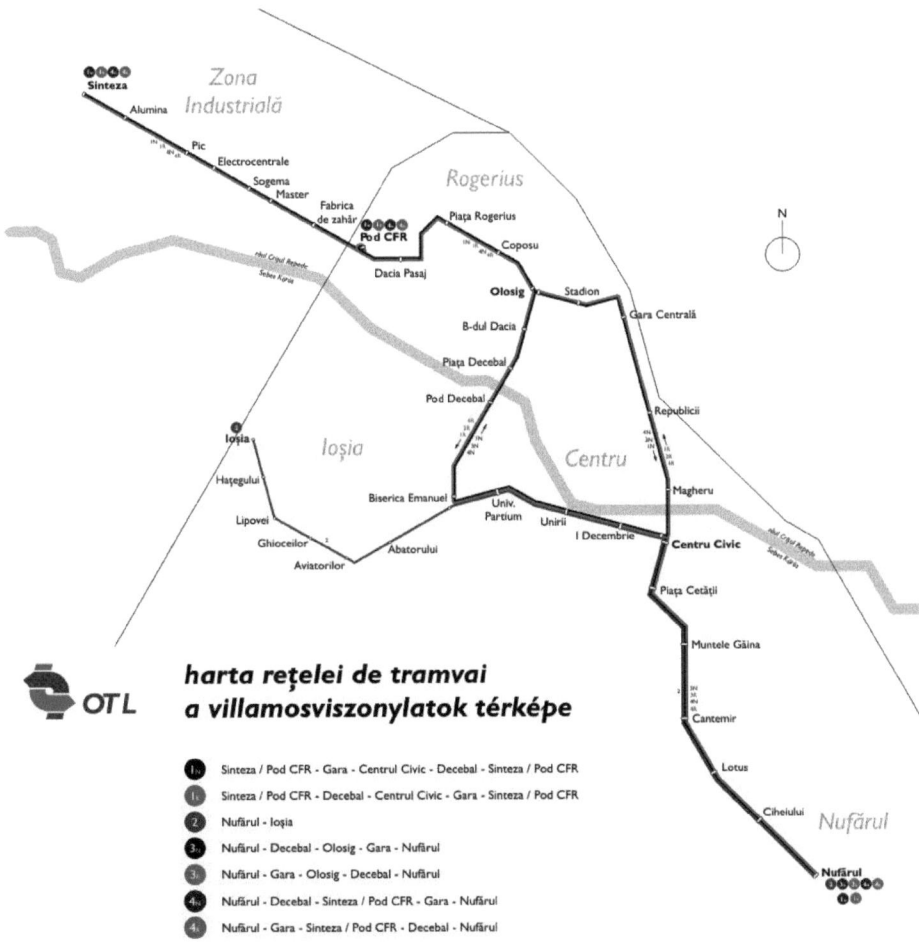

Fig.1.1.a *The Tram network in Oradea municipality (2017)*

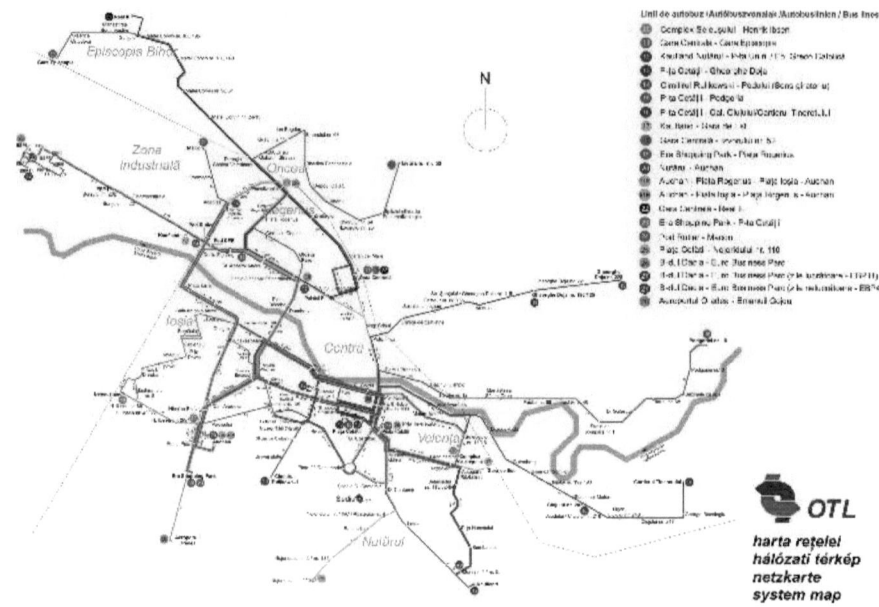

Fig.1.1.b *Bus network in Oradea municipality (2017)*

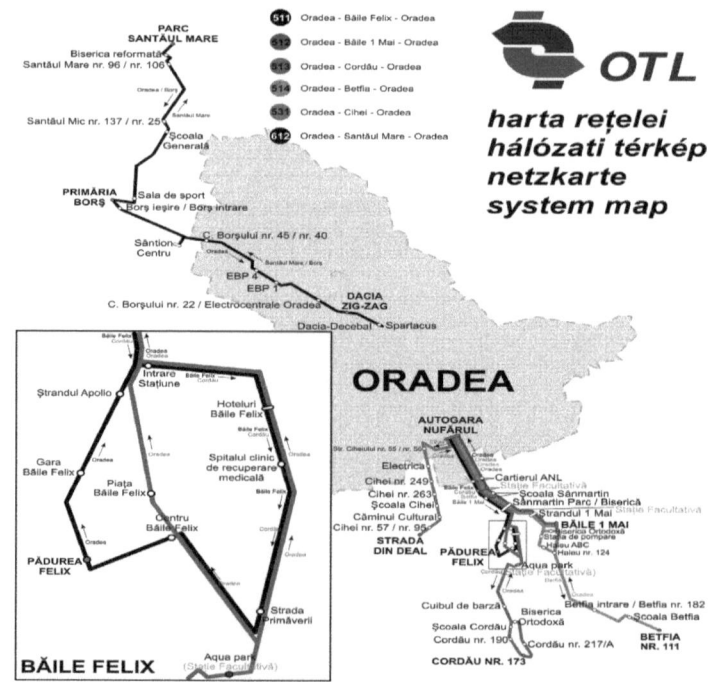

Fig.1.1.c *Bus network in the Oradea's Metropolitan area (2017)*

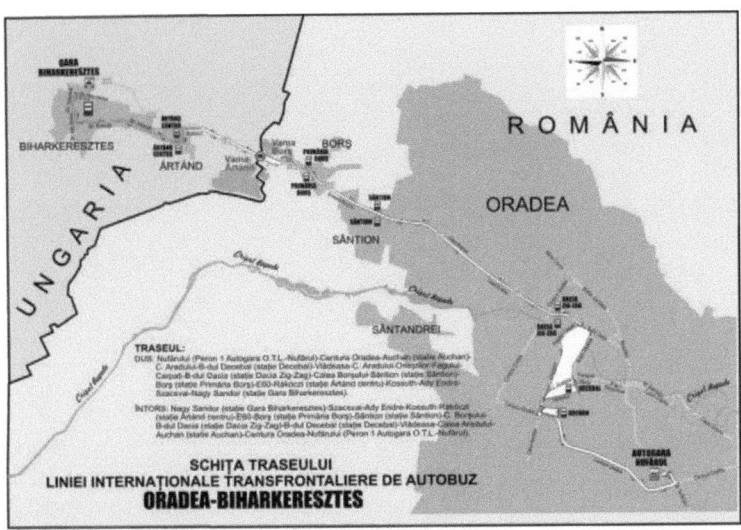

Fig.1.1.d *Cross-border route between Oradea/RO and Biharkeresztes/HU (2017)*

The public transport system analysis challenge how to be compared the tram and bus. The comparison is intended to analyze the tram or the bus per passenger and per kilometer for the use of energy more efficient, cost efficiency in this case is not studied (vehicle and maintenance, human resource cost will collect for economic studies).

1.2.1. Passenger flow analysis in Oradea

Comparing the 2014, 2015 and 2016 average daily data from the main bus lines [5], (e.g. line 12 and 17), we can see two peaks (rush hours) every day, in the morning and in the afternoon. Outside of rush hours the provided vehicle capacity is higher than the necessary. On other bus lines (e.g. line 18 or 19), the number of passengers are not visibly changing. On these lines for the whole day is true the statement that the provided vehicle capacity is higher than the demand.

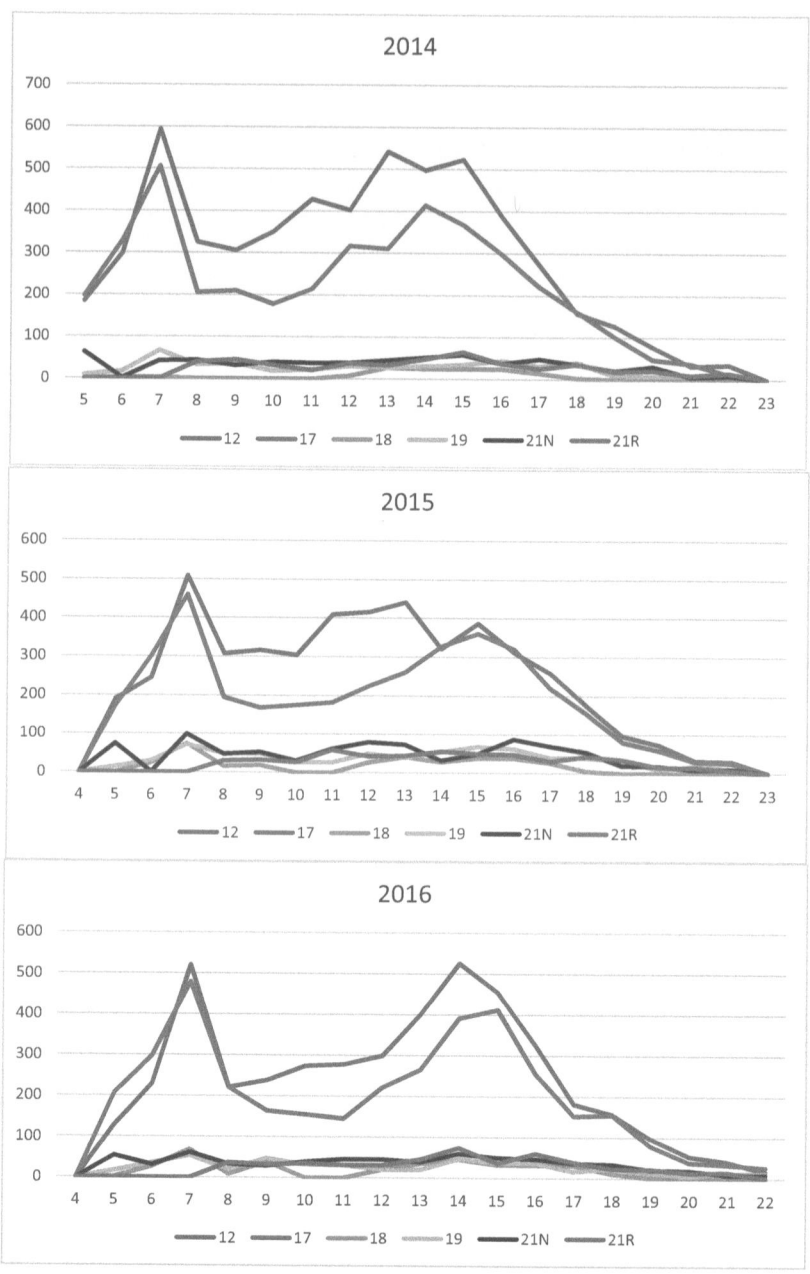

Fig.1.2. *Passenger flow analysis on urban routes*

1.2.2. Fuel and energy consumption data analysis

Comparing the 2014, 2015 and 2016 annual energy consumption and mileage, we can see that with less energy we can provide even higher frequencies on the main bus lines. This method is presented in the strategies used to improve energy efficiency.

Table 1.2.: *Passenger flow, distances, fuel unit consumption*

2014					
Line	Passenge	Consumptic	Distance	l/100km	l/passenger
12	5532	494.8417	1297.15	38.14837914	0.089450777
17	4121	386.1655	991.5	38.94760464	0.093706746
18	126	44.611	119.4	37.36264657	0.354055556
19	470	74.28	204	36.41176471	0.158042553
21N	633	97.74	279.4	34.98210451	0.154407583
21R	482	85.808	245.3	34.98083979	0.178024896

2015					
Line	Passenge	Consumptic	Distance	l/100km	l/passenger
12	4824	501.0005	1374.9	36.43905011	0.103855825
17	3712	357.9819	951.9	37.60709108	0.096439089
18	335	19.6073	118.4	16.56021959	0.058529254
19	690	34.9637	207.4	16.85810029	0.050672029
21N	868	44.6579	274	16.29850365	0.051449194
21R	534	42.9508	254	16.90976378	0.08043221

2016					
Line	Passenge	Consumptic	Distance	l/100km	l/passenger
12	4792	487.4295	1356	35.94612832	0.101717341
17	3667	372.0703	963.9	38.60050835	0.101464494
18	335	63.392	168	37.73333333	0.189229851
19	545	66.684	185.7	35.9095315	0.122355963
21N	784	100.1805	278.6	35.95854271	0.12778125
21R	615	41.7329	254	16.43027559	0.067858374

1.2.4. Easy strategy for energy efficiency: capacity adaptation to real needs

Based on analyzes, the questions are: How can we improve energy-efficiency in public transport to reduce costs? How can public transport become more attractive, "passenger friendly"? Can we improve our provided services and gain a higher profit at the same time? Which public transport systems are more energy-efficient, more economical? Which systems should be a priority, should be developed [6]?

After deeply analysis of passenger flows and correlation with delivered transport capacity, the local public transport company decided in 2014 to optimize the relation between both key performance indicator to achieve small and medium-capacity buses. It was tendered five pieces of KARSAN "Jest", small-capacity bus, with 11 sitting and 10 standing place, and seven piece ISUZU "Novo-city", medium-capacity bus which has 16 sitting and 39 standing places. The KARSAN average fuel consumption is 11 liter/100km, while the ISUZU 16 liter/100km. Compared to, a standard capacity SOLO (12m long, cca.100 passenger capacity) bus consumption of between 31 and 35 liter/100km.

In addition, procurement has increased with five large capacities (18m long, 150 passengers) articulated buses, of which consumption is between 49 and 55 liter/100km.

Fig.1.3. The KARSAN "Jest", small capacity bus (21 passengers)

Fig.1.4.*The ISUZU "Novo-City", medium capacity bus (55 passengers)*

Fig.1.5.*The SOLARIS "Urbino-12", most used capacity bus (107 passengers)*

13

Fig.1.6.*The MAN, high capacity bus (155 passengers)*

Analyzing the passenger flow we could estimate the necessary vehicle capacity we should provide for each line in each part of the day. In 2015 we start optimized the provided vehicle capacity for demand.

Fig.1.7. Average *Consumption/Year/Line*

As we can see on main bus lines, such as the line 12 or 17 we have achieved savings, despite the fact that on these lines, at rush hours, we used as well the large capacity, articulated buses with high consumption data. However, medium and small capacity buses used in off-peak hours not just compensate the higher fuel consumption, but also in comparison with the 2014 data may be noticed a small decline. On the less frequented lines, as line 18 or 19 we managed from 2015 to reduce the fuel consumption per kilometer to half compared to 2014 fuel consumption. The delivered vehicle capacity optimization leads not only to energy-efficiency and economic benefits, but also an increase in the comfort level of passengers on main bus lines by using the large capacity buses during rush hours.

Fig.1.8.*VOLVO Vest, solo bus (105 passengers)*

In 2016 the company decided to procure some "solo", "second-hand" buses with a better thermal comfort to increase the rate of comfort and satisfaction of passengers, decision was to buy 14 pieces of Volvo Vest buses. These have low amounts of run kilometers and well maintained buses offer better conditions for the passengers. The passenger capacity is ~100 passenger and they have an average fuel consumption of 34 liter/100km. Introducing this buses on lines where the passenger flow justified it the fuel consumption rate has increased. On Table 2 we can see this on lines 18, 19, 21N. On line 21R where we kept the small and medium capacity buses we have the same consumption in 2016 like in 2015, almost half of the consumption in 2014. The management decided to use more the small and medium capacity buses on suburban routes in the metropolitan area and on the "cross-border" route. We can see the impact of the relocation of this kind of buses on Table 3. On urban line 12 we can see a constant improvement; on 18 and 21N the effect of using small and medium capacity buses in 2015; on line 21R we can see the impact of using small and medium capacity buses from 2015 and finally the impact of introducing small and medium capacity buses on suburban lines 511 and 512.

Table 1.3.: *Passenger flow, distances and fuel unit consumption*

Year	Line	Passengers	Consumption	Distance	l/100km	l/passenger
2014	12	5532	494.8417	1297.15	38.1483791	0.08945078
2015	12	4824	501.0005	1374.9	36.4390501	0.10385583
2016	12	4792	487.4295	1356	35.9461283	0.10171734
2014	18	126	44.611	119.4	37.3626466	0.35405556
2015	18	335	19.6073	118.4	16.5602196	0.05852925
2016	18	335	63.392	168	37.7333333	0.18922985
2014	21N	633	97.74	279.4	34.9821045	0.15440758
2015	21N	868	44.6579	274	16.2985036	0.05144919
2016	21N	784	100.1805	278.6	35.9585427	0.12778125
2014	21R	482	85.808	245.3	34.9808398	0.1780249
2015	21R	534	42.9508	254	16.9097638	0.08043221
2016	21R	615	41.7329	254	16.4302756	0.06785837
2014	511 FELIX	785	199.593	696.2	28.668917	0.2542586
2015	511 FELIX	411	151.261	528.5	28.6208136	0.36803163
2016	511 FELIX	353	108.223	527.4	20.5200986	0.30658074
2014	512 1MAI	697	187.086	604.2	30.9642502	0.26841607
2015	512 1MAI	624	121.73	408.6	29.7919726	0.19508013
2016	512 1MAI	585	73.5646	461.7	15.93342	0.12575145

1.3.   Tram and bus system energy efficiency comparison

To be able to compare the electricity and liquid fuel consumption data, we need a common denominator. This will be the above-mentioned oil equivalent TOE. The analyzed periods are: 11/10/2014 - 11/16/2014; 11/09/2015 - 11/15/2015 and 11/07/2016 – 11/13/2016. As the tram network covers the main transport axes, most passengers are using this kind of transport system. It is evident that in the test period, approximately two-thirds of passengers use the tram and just one-third the bus.

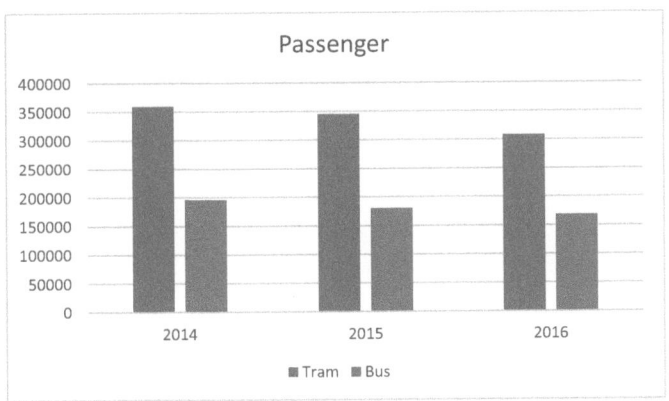

Fig.1.9._Comparison of passenger flow on different transport modes_

Comparing the two public transport system we can see that the distance traveled is inversely proportional to the number of passengers carried. The mileage data includes data from the metropolitan area lines as well.

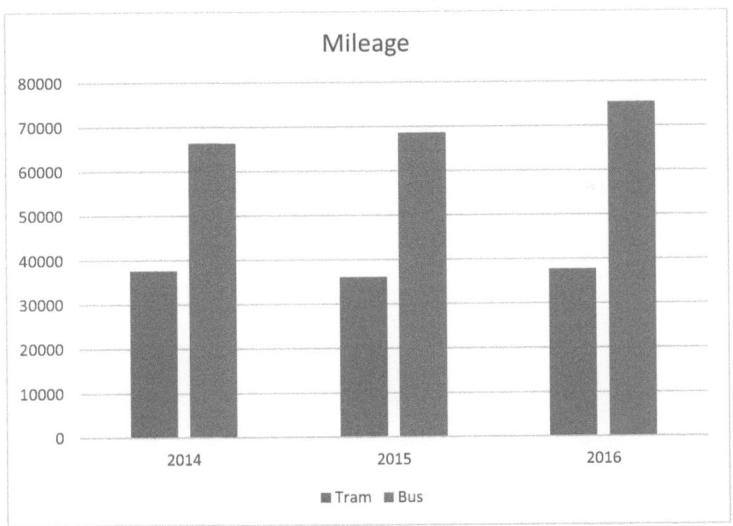

Fig.1.10._Mileage of trams and buses per year_

Value (in toe) of energy consumption for Tram system shows almost half of bus system indicator. The applied strategy (to adapt transport capacity to the real passenger flow) reduced the energy consumption of bus system.

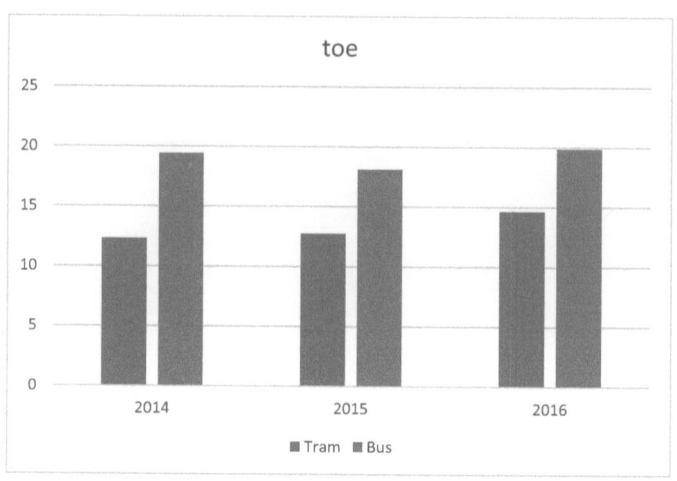

Fig.1.11._Energy consumption calculated in toe_

If we compare the energy consumption (in toe) to the number of passengers carried, the tram systems energy-efficiency is very obvious. Approximately one-third of energy consumption of the bus system per ten thousand of passenger in 2014 the tram system used 0.34 toe of energy, while the bus system is 0.99 toe, in 2015 energy consumption is similar 0.37 toe to 1.00 toe. In 2016 the covered distance has been increased while the carried passengers has been decreased, due to this the energy consumption per passenger increased as well to 0.47 toe and 1.18 toe.

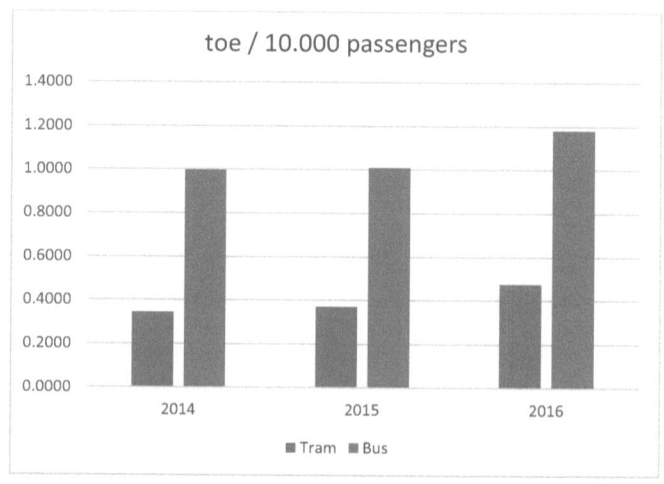

Fig.1.12._Energy Consumption per 10.000 passengers_

The only examined indicator in which the bus system looks effective than the tram system is the energy consumption per covered distances.

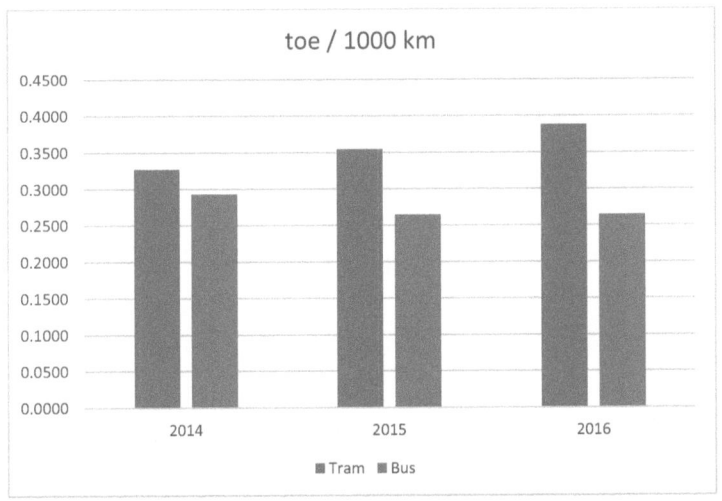

Fig.1.13.*Energy consumption for 1.000 km*

1.4. The evaluation of the "cumulative" energy efficiency indicator

Because between the analyzed periods, the transport company has bought only buses to increase the level of energy efficiency, energy efficiency analysis will only reflect this kind of public transport system. In the future, when we will modernize the electric traction of trams will lead to lower energy consumption and we could compare the evolution of the two systems.

To analyze the evolution of energy efficiency we will use the ODEX indicators. The ODEX indicator system was designed and developed by an international consortium led by the French company ENERDATA in Grenoble under the SAVE program of the European Commission and continued under the Intelligent Energy for Europe Program. According to Directive 2006/32/EC, ODEX indicators will be used to implement the complex reporting methodology for the implementation of Energy Efficiency Action Plans. ODEX indicators are percentage indicators; A reference year is chosen for which the value of the given indicator is 100% and its values are calculated for other years; if those values are greater than 100% means that energy efficiency has decreased (energy consumption to achieve a given useful effect has increased); If those values are less than 100%, it means that energy efficiency has improved. ODEX indicators do not describe the level of energy efficiency (low or high) but the evolution of energy efficiency (positive or negative trend). ODEX indicators describe the dynamics of the process and not its state at some point. They do not allow comparisons for a given year between two countries, two consumer sectors but only comparisons for different years for the same consumer sector. The ODEX indicator is calculated for a specific sector or level of aggregation (national economy, manufacturing, transport, final consumer etc). The ODEX indicator at a specific sector or level of aggregation (national economy, economic sector, final consumer, etc.) is calculated as a weighted average of lower energy efficiency indicators. In practice, each work group can set its own analysis structure according to its specific characteristics and the available primary data [7].

The ODEX indicator (IOD) is by definition the percentage of the ratio between:

$I_{(t)}$ - energy consumption in year t, with which a certain quantity of goods or services has been obtained; and

$I_{(t0)}$ - the energy consumption that would have been needed in year t if the energy efficiency was the one in year t_0 (a previous year conventionally chosen) [7].

$$IOD(t) = \frac{I(t)}{I(t0)} \times 100 \qquad (1.1)$$

We will analyze the lines where the delivered transport capacity was adapted to the requirements of the passenger flow.

Table 4.a *ODEX indicators on the analyzed lines focusing on the covered distance*

ODEX - mileage			
Line	2014(t0)	2015(t1)	2016(t2)
12	100	95.5193	98.6473
18	100	44.3229	227.855
21N	100	46.591	220.625
21R	100	48.3401	97.1644
511 FELIX	100	99.8322	71.6964
512 1MAI	100	96.2141	53.4823

Table 4.b *ODEX indicators on the analyzed lines focusing on passenger flow*

ODEX - passengers			
Line	2014(t0)	2015(t1)	2016(t2)
12	100	116.104	97.9409
18	100	16.5311	323.308
21N	100	33.3204	248.364
21R	100	45.1803	84.3672
511 FELIX	100	144.747	83.3028
512 1MAI	100	72.6783	64.4614

1.5. Conclusion of survey analysis

It is important that the mileage of the bus system is approximately twice of the covered distance by the tram system, but the energy consumption per covered distance of the tram system is not significantly higher. Notice that in 2015 the bus system energy consumption indicator is better than in 2014 despite the fact that that the covered distances are increased. As we can see in 2016 starting to operate the 14 pices of high capacity VOLVO bus with AC and better thermal confort, replacing medium capacity buses on some routes (21N is a relevant example), the energy consumption has increased, but it solved the complain of crowded trips. The passenger rate on bus decreased from 90-95% to an average not more than 55%, passengers been satisfied by the confort of new buses.

By analyzing the ODEX indicators, we can see the evolution of the energy efficiency indicators on the lines where the delivered transport capacity was adapted to the requirements of the passenger flow at certain intervals of the day. We can see the the impact of using small and medium capacity buses on lines where they have been introduced.

In the future, when the company will purchase trams with lower transport capacity and modern traction, or upgrading the existing tram park with new traction, we will be able to calculate the ODEX of the electric public transport system. Then we will be able to compare the evolution of the two transport systems.

Observation:

This study was made on frameworks of MECHATRONICA Laboratory/Oradea Transport Local SA (having as co-author Eng. László LIBOR, informatics engineer at same company), as preliminary paper was presented during Conference "CIE-2017", 23rd Regional Energy Forum, 08 - 09 June 2017, organized by University of Oradea in Băile Felix/ROMANIA and published at "Towards a humane city" - 6th International Conference, organized by Department for Traffic Engineering, Faculty of Technical Sciences, University of Novi Sad, 26 - 27 October 2017.

Study 2 - The urban electric bus, a sustainable solution to increase energy efficiency of public transport and reduce atmospheric pollution in the cities

2.1. About energy consumption in transport sector

2.1.1. World climate change

Climate change is a cruel reality, news about the disasters are daily. Although the world's leading heads of state and government agreed to cooperate in the development, application and availability of climate-friendly technologies for developing countries, however, there are blockages, it is enough to refer to the United States 2017's "turnaround", hopeful the temporary rejection of a treaty accepted by the majority will not trigger an avalanche reaction.

We know there are many environmentally friendly alternatives, this is why urgent action is needed, and climate protection must be one of the foundations for development goals. Regional and urban mobility is a key condition for economic and social development and the reduction in poverty. The transport sector therefore plays a key role in the development and maintenance of the economic and social foundations of both developed and developing countries.

2.1.2. Transport's energy consumption and air pollution

The continuous growth in the transport sector has increased concerns about the economic costs of energy supply as well as the impact on the environment. In the EU, the road transport sector is responsible for 26% of final energy consumption and 24% of CO_2 emissions (2007). Energy use and emissions from the road sector continue to grow around 1.5 - 2 % per year [8].

The 2016 Report of the International Energy Agency (IEA) [9] found that the share of transport in world oil consumption has increased significantly over the few last decades. Fig.2.1. and fig.2.2. show the transport sector energy needs and evolution comparing 1973 to 2014. If we take the absolute equivalent energy values the increasing electricity using succeed to replace the classical fuels.

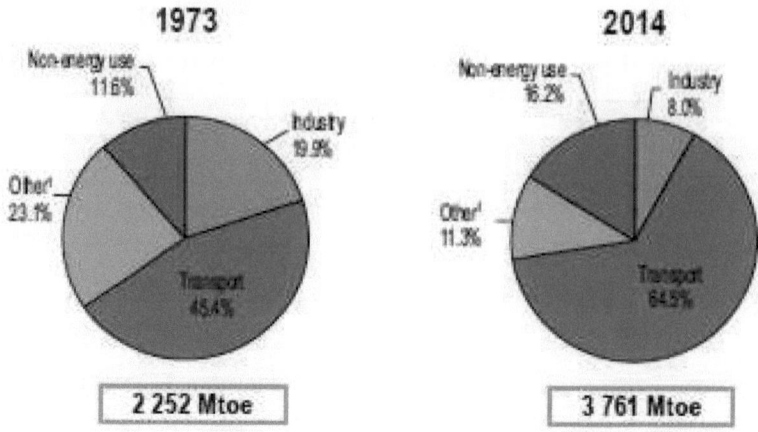

1. Includes agriculture, commercial and public services, residential, and non-specified other.

Fig.2.1.*Oil consumption by sector* (source: IEA)

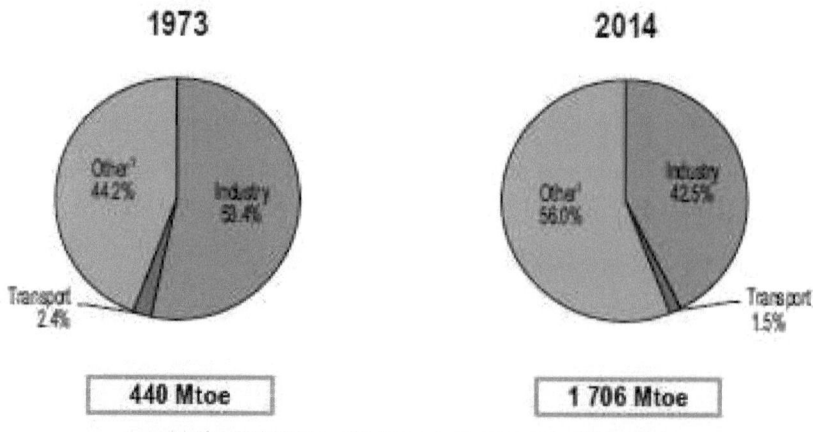

1. Includes agriculture, commercial and public services, residential, and non-specified other.

Fig.2.2.*Electric energy consumption by sector*(source: IEA)

The comparison marks the first oil crisis as a milestone, according to data from 1973, the share of transport accounted for 45.4% oil consumption in 1973, nowadays nearly two-thirds of crude consumption is in the transport sector, in 2014 the rate was 64.5%.

Although the share of electricity used in transport declined from 2.8% in 1973 to 1.5% in 2014, but if we take absolute value, it is more than twice the amount of electricity consumed since the 1970's. Taking into account the significant advancement of energy efficiency in electric propulsion techniques, it eventually brings about more than twice the vehicle ratio, i.e. the share of electric vehicles has increased.

2.2. Importance of the public transport

2.2.1. Urban mobility

The UITP (International Public Transport Association), based in Brussels, jointly with the EU Commission, strongly recommends that European climate change and energy efficiency policies explicitly address the issue of urban mobility. One of the most important measures proposed is the development of a strategy to promote modal shift and high quality public transport systems [10]. One element of this is the need to develop sustainable urban mobility/transport plans across Europe for cities with more than 100.000 inhabitants (governed by the HG 191/2013 in Romania). Supporting investments in transports, addressing high energy efficiency and low greenhouse gas emissions as priority criteria accentuate the vital role of public transport. Unfortunately, it is often considered to be a too costly endeavor improving energy efficiency and the environment (reduced emissions).

Electric drives are able to solve the challenges of public transport:

- Reducing local pollution, whether comparing other used engine is gas, liquid or solid,
- The significant reduction of the CO_2 emissions,
- Compliance with lower noise emission levels,
- Future reduction of the oil based fuels production and possible inflation of oil prices,
- The possible decreasing of internal combustion engines used in public transport will not affect the development and increasing of urban mobility.

2.2.2. About electric drive vehicles

The best solutions for a high passenger flow with more than a few thousand trips per hour are the electric powered systems, whether it is metro or tram, trolley or the electric bus. One of the major challenges of the future is what kind of electric energy we can supply to the vehicles and from which sources, what autonomy it is given by the energy storage, how long it takes to recharge it and with what kind of standard.

The main advantage of electric propulsion is the power-torque characteristics. Typical Torque Curve for 1PV5 type Siemens AC Induction Motor [11] presented in Fig.2.3. indicate a high efficiency solution. Fig.2.4. explains energy resources of buses in sketched graphic design. Energy can also be recharged on a vehicle on-the go (generated by internal combustion engine, from outside electric network as direct connection or as inductive charging), or on-stopped (charge of battery or supercapacitor). Recovery of brake energy is common in today's technology, can be reused immediately by other vehicles or reinjected back into the electric energy network. The flywheel method is a mechanical facility conserve part of recovered energy. Many concepts have main scope to achieve the most efficient torque for electric vehicles.

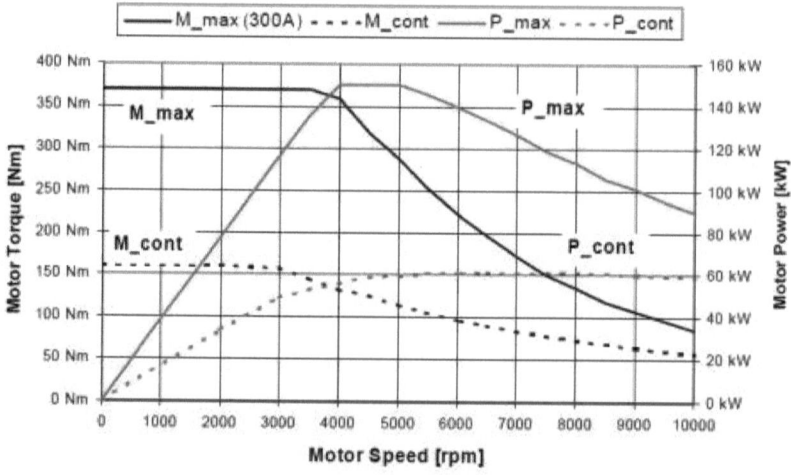

(source: siemens.com)
Fig.2.3. *Example of efficient electric motor characteristic*
(power and torque curves for an 1PV5 Siemens motor) (source: siemens.com)

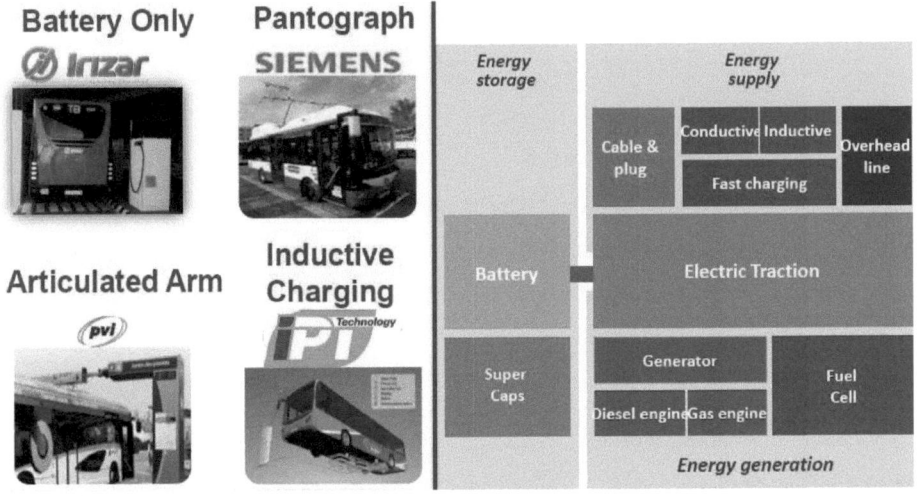

Fig.2.4. *Charging and recharging methodes* (source: ZeEUS)

2.2.3. Where are the electric buses?

Assessing the present fleet in Europe at 2016 [12], oil-dependent transport systems cannot be replaced in a short time at a low cost, it needs substantial financing and it is an expensive undertaking for any Government.

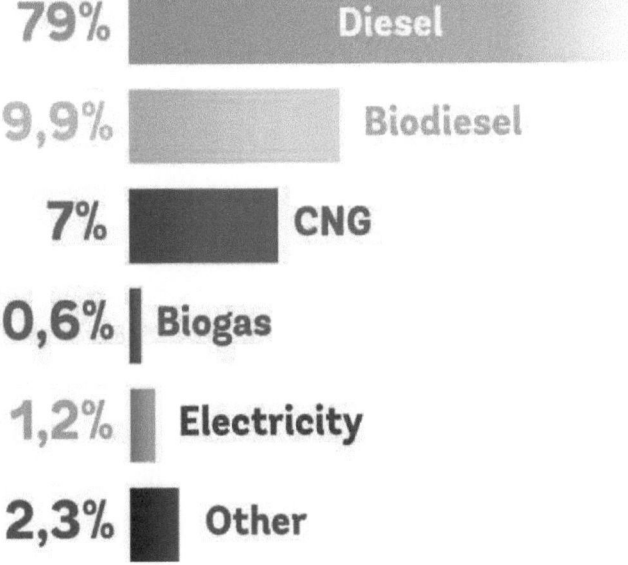

Fig.2.5. *Bus engines in Europe - 2016 (source: EBSF2)*

Knowing the vehicle prices, based on the standard 12 m long urban "solo" bus type, which depends of dotation is starting from 150,000 to over 300,000 euros in diesel and increasing to several hundred thousand euros for the hybrids, not to mention the price of a "super computerized", ultramodern electric bus which is up to a million euros. As we can see there is a very broad price spectrum. Prices depend on the type, brand, equipment, quality and maintenance services agreements but in the end the authority's budget will determine which manufacturer to sign a contract with.

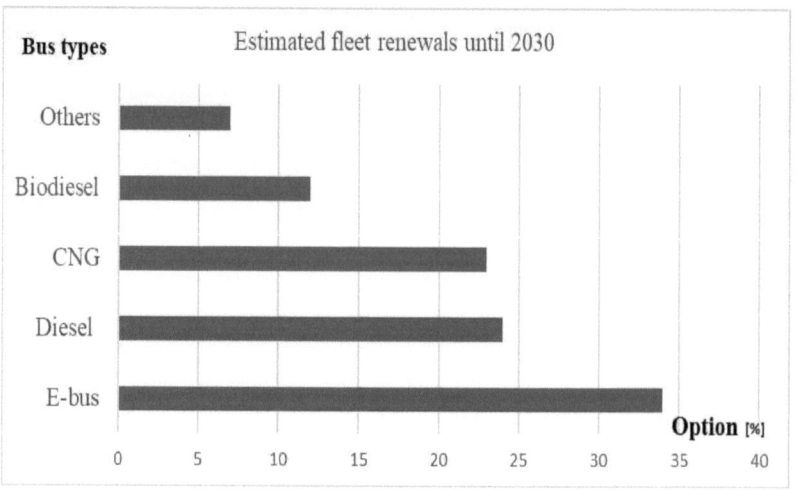

Fig.2.6. *Respondents distribution according to future plans to change propulsion system ratio* (source: EBSF2)

As a result of a general survey in EU Project [12], shown in fig.2.6., we know that today the electric buses represent only 1% of the surveyed fleet and more than 34% of the respondents want to change towards more electric solutions, mainly hybrids and fully electric with batteries, as shown in fig.6. Significant patterns are also evident for more CNG (23%), biodiesel and biogas (12%). The total proportion of other fuels (hydrogen, bioethanol, etc.) is about 9%.

Energy optimization and environmental friendliness does not characterize today's inventory of bus types, although the automotive industry offers more and more technical solutions. The following areas have a special priority [13]:

• Energy recovery by regenerative breaking, heat recovery from exhaust system and energy or heat recovery from heating or cooling of the hybrid batteries

- Weight reduction using lighter more durable composite materials and redeveloping the structure of the vehicles using these new technologies

- Optimizing the traction by eliminating friction energy losses, adding more start-stop functions and using more efficient drive chains.

2.4. The electric bus (e-bus)

2.4.1. Electric and Hybrid over Diesel

Fully electric powered buses are widespread and operate in large daily traffic areas in several large cities. Their polluting gas emissions are zero (at the site of use!) and the noise emission is lower than diesel buses. However, greenhouse gas emissions continue to depend largely on the production method of the electricity used, and the manufacturing process of the batteries.

In the long term, for the achievement of the EU's 2050 targets [10], electric buses show the best prospects, which might be supplemented by hydrogen fuel cell buses on longer routes. This is due to the high energy efficiency of the entire electric drive, which together with the potential of using renewable energy production methods can result in low greenhouse gas emissions and energy consumptions altogether.

2.4.2. European Union's recommendations

The basic legislation regarding the four major transport sectors (air, rail, road and water) is the – "White Paper on Transport/Roadmap to a Single European Transport Area" (COM 2011/0144). The European Commission has issued a mandate "M-533" calling on CEN/CENELEC (the EU Commission for Standardization) to propose a standard for the classification of electric vehicles by the end of 2019 [10].

"Key Performance Indicator" - KPI are utilized for a multitude of features that make the different characteristics comparable. According to the UITP recommendation, within the framework of the ZeEUS project, 187 different performance attributes have been selected and mapped. After the analysis of these attributes they have reduced the number to 18 main KPI indicators, so the comparison will be easier and transparent [14].

2.4.3. Infrastructure of chargers

When switching to electric buses, conflicts between parallel or competing technical solutions must be avoided. From the efficiencies point of views it should be possible to connect as many vehicles as possible to the same charging apparatus or charging station. Existing tram or trolleybuses networks can be an excellent basis for the electric charging system [13].

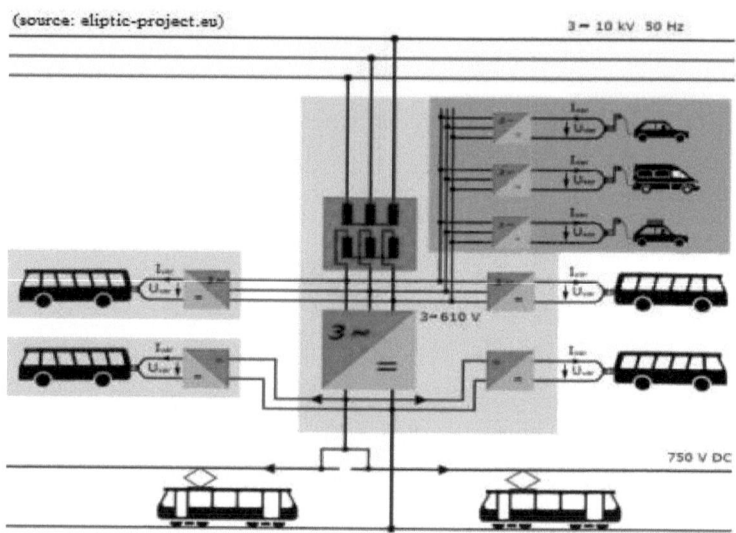

Fig.2.7. *Power stations as optimal charger points* (source: ELIPTIC)

Fig.2.7. explains how a classic electric power supply network (the overhead line system work in parallel), can be easily converted to charging units of the electric bus fleet and other electric cars.

2.4.4. Survey for energy requirement

The capacity of the 600 V batteries and the charging cycle of the electric bus can be determined [11] by taking into account the available charging time and the amount of charging power required.

In order to select an electric vehicle, it is necessary to specify exactly what the customer desires. It must be decided whether environmental issues and other people-centered considerations are solid enough and acceptable for investing in a high-priced electric bus.

There are several methods to calculate and approximate technical parameters of charging equipment. One of them is a graphical approach [11] as a histogram show dependency of the batteries capacity, weight and volume, power needs for bus, charging methods and the available time for that.

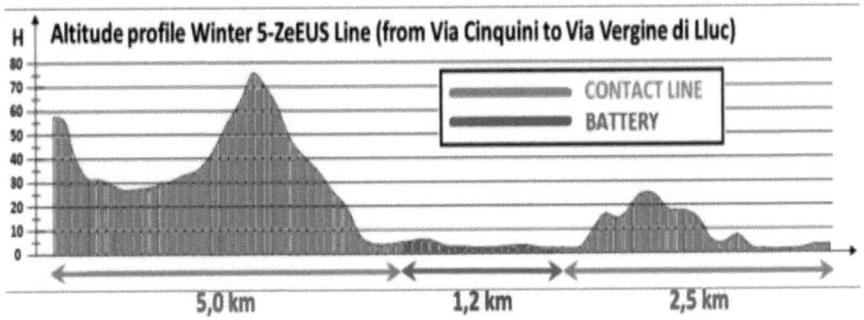

Fig.2.8. *Elevation on Route 5 – Winter* (source: ZeEUS)

Fig.2.8. and Fig.2.9. shows a bus route, namely Line 5 of the ZeEUS partner, Cagliari in Italy. Three main sections can be distinguished, two green areas (2.5 and 4.0 km) on a hilly district and a red section on a flat-lane. On the green sides overhead contact lines are needed for trolleybus operation. On the red area where City Authorities do not allow overhead lines it is possible and mandatory to run on battery mode.

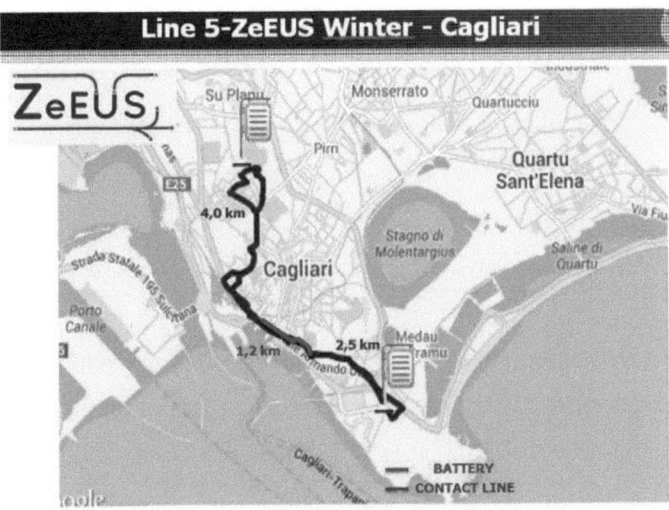

Fig.2.9. *Cagliari – Route 5 – Winter on map* (source: ZeEUS)

Knowing the terrain condition is it possible to determine the vehicles configuration, such as energy storage and energy sources.

Charging can be done in four separate ways. Four modes of charge are used: battery cables, pantograph, armed connector and inductive charge.

The energy structure of vehicles can be analyzed with a SANKEY diagram. In the Fig.2.10. and Fig.2.11. shows the SANKEY diagram for diesel and electric busses. The reduction of losses is one of the main priorities for increasing the efficiency of used energy, but the reversible energies are also important elements of operation [1], [2].

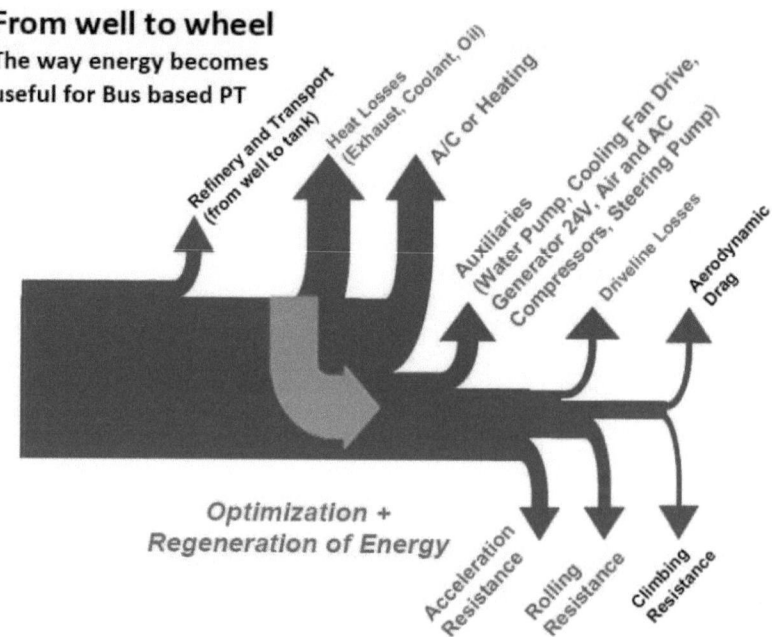

Fig.2.10. *Sankey diagram for Diesel buses* (source: *ELIPTIC*)

The thickness of the arrows indicates the ratio of the energy invested. The green arrow means the energy recycling of the internal combustion engine: heat recovery and reduced heat loss, proportionally greater drive force.

In order to optimize the efficiency of electric-powered vehicles, it is of utmost importance to recycle the recovered energy directly and efficiently in the electrical grid (green return arrow thickness is the amount of recovered energy).

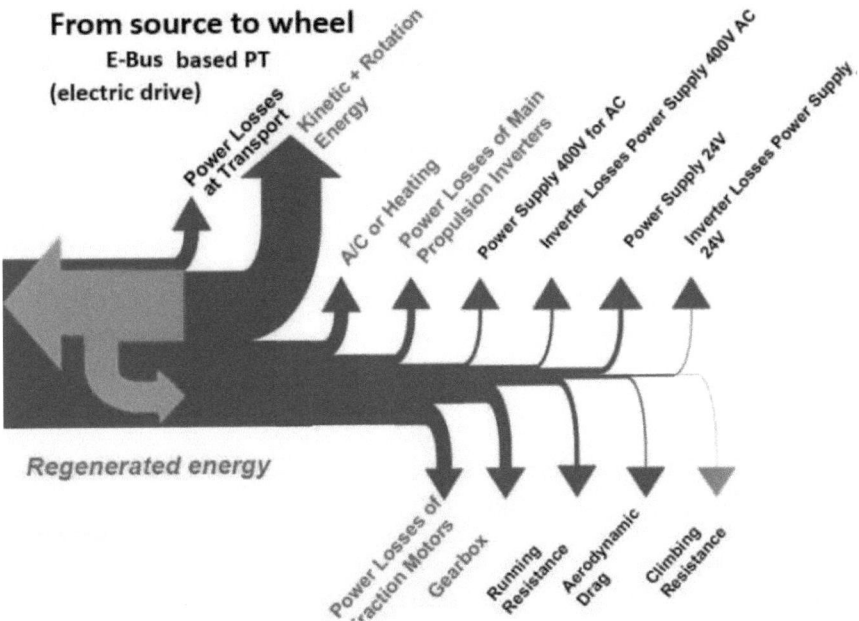

Fig.2.11. *Sankey diagram for Electric buses* *(source: ELIPTIC)*

2.5. Chargers for e-buses

2.5.1. Galvanic connection

Within the framework of the ZeEUS project [14], for the bus tested at the Bonn SWBV, FRAUNHOFER company recommends the following measuring equipment: voltage converters LEM LV 100/SP47, current converters LEM ITN 600-S, GPS receiver, Compact-Rio data logger. This equipment has the possibility to monitor and register instantaneous power at a given GPS position. Territorial mapping of energy consumption can be used to calculate the route energy gain. Recharging durations are particularly important where charging is used at intermediate stops. At such a bus-stop there is a limited amount of time which can be spent for recharging. Within the test, direct energy consumption (EC_{Bus}) measurements are performed on a bus, calculated by (2.1.), which can be compared to the consumption values measured at the charging stations.

$$EC_{Bus} = \int_0^t (U_t \times I_t) \qquad (2.1.)$$

Measuring at the same time the charge energy, knowing the initial charging ratio, after running the replenished electric energy to the same point, EC_{charge} energy value can be obtained (as used full to full option). By allocating the charge energy to the energy consumed on the bus (useful, effective), C_{EC} energy efficiency coefficient can be calculated by (2.2.):

$$C_{EC} = \frac{EC_{Bus}}{EC_{charge}} x100 \qquad [\%] \quad (2.2.)$$

2.5.2. Equipments, connector standards

The world-wide standard **IEC62196/2011** recommends the following connectors:

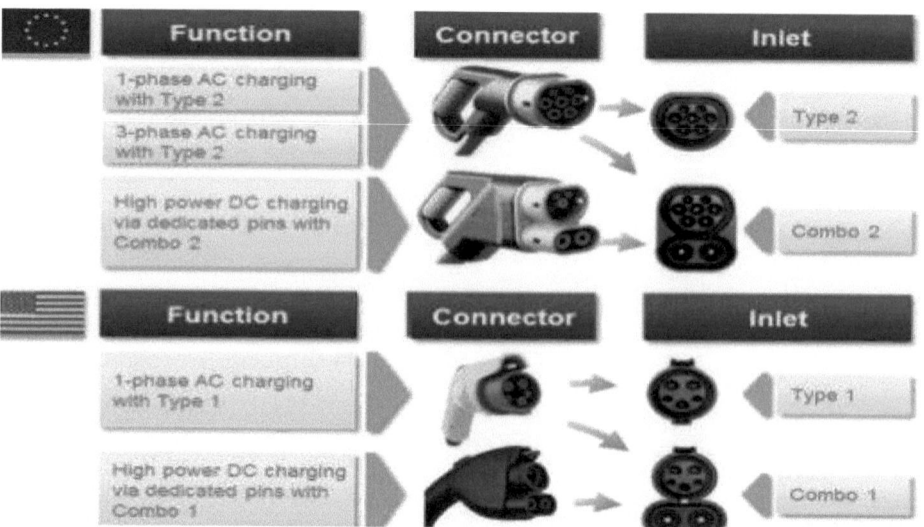

Fig.2.12. *Charger connector and inlet standards* (source: IEC62196)

- For American systems - **CCS 1** (but also accepted for CCS 2) for electric buses in North America
- For European systems - **CCS 2** (Combined Charging System) as a charging interface for Europe's electric bus manufacturers

Synergies between car and bus solutions are possible. The fig.12 presents industry standards for chargers and inlets both in the United States and Europe, as recommended solutions.

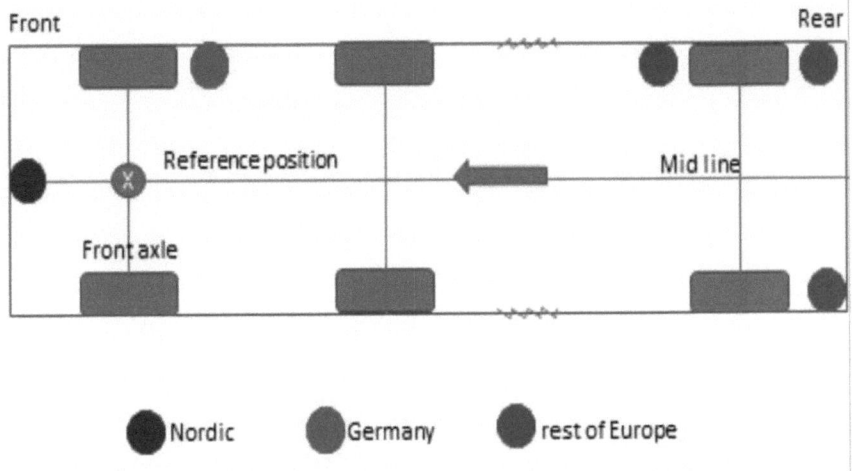

Fig.2.13. *Recommended positions for chargers* (source: ELIPTIC)

Positioning the charging connector is essential to eliminate unwanted heat losses due the high charging current. Fig.2.13. shows recommended charger positions for electric busses.

2.6. Case studies

2.6.1. Belgrade: Eko 1, the E-Bus Route

The new route <u>EKO 1</u> is shown in Fig.2.14. [15] link between the old and new Belgrade that crosses the crowded City Center using HIGER A6L E-Buses.

Fig.2.14. *Route EKO 1 inside through the downtown* (source: Chariot Ltd)

Fig.2.15. *The Electric Bus model "HIGER A6L" in Belgrade* (source: Chariot Ltd)

Analysis of the charge indicators by graphical representation of the measurement statistics defines the energy consumption range between 0.85 and 1.23 kWh / km, which varies according to field and operational impacts. The terrain features and outdoor weather (air conditioning or heating) determines the highest energy consumption, but also the traffic factors (crowded downtown, traffic jams, etc.). The volume of passenger flow has a strong influence itself as an important factor on energy usage. The daily average consumption noted in Table 2.1 HIGER type Chinese buses have been running on the Route EKO 1 since the fall of 2016 (similar run in other places: Sofia/Bulgaria, Graz/Austria and Tel Aviv/Israel).

Table 2.1. Daily statistics in Belgrad: average consumptions

E-bus No	02-09-2016 (30°C) AC-on	06-09-2016 (24°C) AC-off	20-09-2016 (20°C) AC-off
2101	1.26	0	1.04
2102	1.14	1.01	0.89
2103	1.13	1.06	0.97
2104	1.14	0.97	0
2105	0.97	0.88	1.00
Average	**1.128**	**0.98**	**0.975**

Table 2.2. Diesel and Electric bus comparison

Averages / Bus type	E-bus „Higer A6L"	EuroVI Diesel bus
L- yearly average operation milage (km)	80,000	
E_u- average energy consumption (kWh/km)	0.99	-
ENERGY consumption - total (kWh/year)	79,200	-
Ce_u - electric energy unit price (Euro/kWh)	0.07	-
F_u- average fuel cons-umption (liter/100 km)	-	44
FUEL consumption - total (liter/year)	-	35,200
Cf_u- average fuel unit price (Euro/liter)	-	1.17
ANNUAL ENERGY TOTAL COST (Euro/year)	5,544	41,184
CE_{dif} – annual energy cost difference (Euro/year)	**35,640 euros**	

Taking into account the market average prices, a low price electric powered bus (E1): ~400,000 euros, initial cost of ownership, without any charger infrastructure, equipment. The operating, maintenance and energy costs must be calculated for the whole life cycle cost. To this end, it is worth adding ~150,000 euros, the battery replacement (A1), manufacturers openly admit that besides today's technologies there is a guarantee only of four years for today's batteries. A similar size (12 m) classic Euro VI bus (D1) is available at around 200,000 euros. Operation yearly values were in included in Table 2.2., which help analyze difference between the Diesel and Electric Bus.

Equation (2.3.) calculates the annual cost differences (CE_{dif}) between the two buses:

$$CE_{dif} = [\left(\frac{L \times F_u}{100}\right) \times Cf_u] - (L \times E_u \times Ce_u) \quad (2.3.)$$

Applying average values contained in Table 2.2., the yearly difference is in the amount of 35,640 euros.

$$T_{inv} = \frac{(E1+A1)-D1}{CE_{dif}} \qquad (2.4.)$$

The Return of Investment (ROI) time (T_{inv}) is calculated by (2.4.) taking into account the price difference between the E-Bus and diesel (550,000 euros and 200,000 euros) divided by 35,640 euros per year. The end result is nearly ten years of amortization, more precisely: T_{inv} = **9.82 years**.

An electric bus has an average life expectancy of 8 years, as such it is not a desired investment for economic reasons!

2.6.2. Oradea: Measurements and test results

In February 2015, a Chinese-made BYD electric bus was presented to the press and trialed by specialists from the public transport company OTL SA (Oradea Transport Local SA- www.otl.ro). The same year, early November, the Czech made SOR E-bus was tested for a period of two weeks on different routes. Namely Route 14 (urban cross city) and Cross-border route to Biharkeresztes/Hungary (long range more than 40 km).

The test run and the measurements were designed to prepare, for the acquisition of one or more electric buses.

The measured values with FLUKE 345 were analyzed in a statistic Table 2.3., energy efficiency and power factors provided satisfactory results.

Table 2.3. Daily statistics in Oradea: average consumptions

Date [year 2015]	Distance [km/day]	Charging [h]	Average [kWh/km]
1-Nov-2015	80	4.10	1.0625
2-Nov-2015	139	8.30	0.8705
3-Nov-2015	123	7.20	0.8618
4-Nov-2015	47	3.30	0.6383
5-Nov-2015	82	2.10	0.9024
6-Nov-2015	123	9.00	0.9024
7-Nov-2015	47	4.20	0.766
8-Nov-2015	150	3.20	0.9467
9-Nov-2015	113	3.00	1.0177
10-Nov-2015	59	4.40	0.8644
11-Nov-2015	128	3.50	0.8516
12-Nov-2015	144	8.30	0.7639
13-Nov-2015	93	6.10	0.9032
14-Nov-2015	90	5.90	0.8922
15-Nov-2015	79	6.00	0.7835

The values are very similar to the Belgrade example, between 0.6835 and 1.0625 kWh/km, with an average of unit distance energy consumption of 0.8728 kWh/km. For a smaller capacity bus (10 m instead of 12 m) shows very good results for energy efficiency. Knowing the comparing of different Oradea Transport Local SA's networks (buses & trams) energy consumption (in TOE- Tone of Oil Equivalent) reported to the number of passengers carried, the tram systems energy-efficiency is very obvious [16]. The differences are essential, approximately one-third of energy consumption of the Diesel engine based bus system. Per ten thousand of passenger in 2014 the tram system used 0.34 TOE of electricity, while the fuel combustion bus system is 0.99 TOE, in 2015 energy consumption is similar 0.37 TOE to 1.00 TOE (a light increase by changed rules on heating and HVAC time extension).

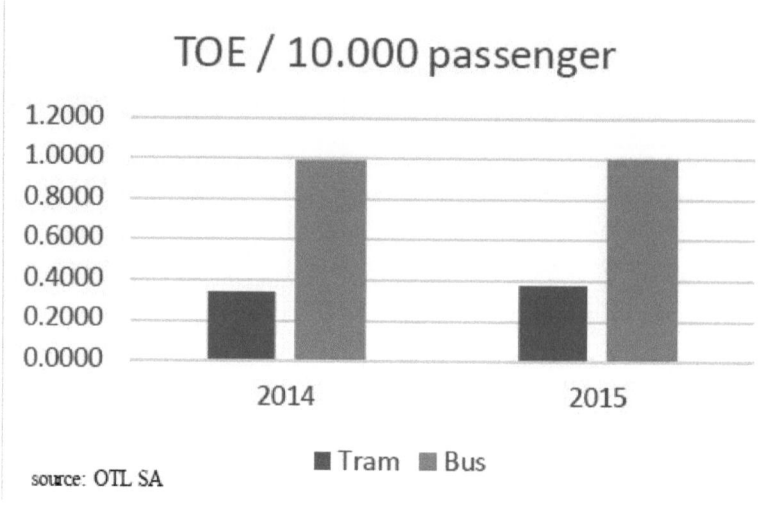

source: OTL SA

Fig.2.16. *Average energy consumption*

per 10.000 passenger during the two year of the test period (source: OTL)

As show in Fig.2.16. the recommendation for increasing energy efficiency are very clear in favor of electric driven public transport, including the electric buses.

2.7. General conclusions about E-Bus utility in Public Transportation

Although this paper is drafted as a technical study, it had to be completed with economic content in order to better understand the usage of E-Buses. There is a lot of money in the field of environmental protection, and governments support more or less the issues. An IMF (International Monetary Fund) study states that support schemes have a serious fiscal, welfare and environmental impact. According to calculations and simulations, with the withdrawal of major post-tax subsidies, globally the Governments revenues would have risen by 3.6% of GDP in 2015. It is important too: the mortality rates caused by air pollution would have been halved [17]. This decision is not made by engineers, but by politicians. Social backlashes, inflation and unemployment must be taken into account when shifting to eco-friendly industries. People need to be trained for new job markets.

The European Commission looks into a cost-effective way to make the European economy more environmentally friendly and energy-efficient. The – "Roadmap for a Low Carbon Economy", CE112 (2011), suggests that by 2050 the EU should reduce emissions by at least 80% below 1990 levels, with milestones being 40% by 2030 and 60% by 2040 [18].

Debate is open, engineers can argue whether the electric bus is environmentally friendly or not! This study had to take into account the whole process of manufacturing such a vehicle, which includes using rare earth materials. End of life costs are also higher for electric busses given the higher hazardous material content. This is the main difference between Life Cycle Cost Management and Total Cost of Ownership, given that the owner should not only take on the costs and profits of the operation, but the vehicle's (or other products) owners are also responsible for the end operation, after-removal treatment costs.

Without energy, our society does not work. Our viable future lies in a much more balanced energy mix, and this will continue to play a role in fossil fuels in the long term, just as in the case of the highly growing proportion of renewable energy sources or the otherwise zero-carbon nuclear power.

Observation:

This study was made on frameworks of MECHATRONICA Laboratory/Oradea Transport Local SA (having as co-author Eng. Botond CSUZI, telecommunication service engineer at ROMATSA SA) paper was presented during Conference "ELECTRIC MOBILITY EV-2017", The 12th edition of the International Conference and Expo Show, 5-6 October 2017, Bucharest/ROMANIA.

REFERENCES:

[1] I. FELEA, I. CSUZI, E. BARLA: *Modelling and Assessment the Energy Performance of an Urban Transport System with Electric Drives*, Revue Traffic & Transportation "PROMET", Vol.25/2013, p. 495 – 506, Zagreb / Croatia

[2] I. CSUZI: *Contribuții la evaluarea și optimizarea performanțelor energetice și de disponibilitate ale sistemului de tracțiune electrică urbană-RO;* (EN- *Contributions to the evaluation and optimization of the energy and availability of the urban electric traction system)*, Ph.D. Thesys (prof.dr.ing. Ioan FELEA), 2011, Universitatea Oradea/ROMANIA

[3] I. FELEA, I. CSUZI, E. SILAGHI: *Global indicators used for the availability performance evaluation of an urban transport system using electrically driven trams*, Review "Buletinul AGIR", 2012, p. 447 – 455, Bucharest / Romania

[4] ***: International Energy Agency Calculator, www.iea.org

[5] CSUZI I.: *Hogyan működik a nagyváradi közösségi közlekedés – elektronikus flotta menedzsment és kártyás jegyrendszer-HU; (EN- How it works the public transport in Oradea - fleet management and e-ticketing system)*, Conference "T-SYSTEMS TEMATIK 2015" Magyarországi Közösségi Közlekedési Konferencia, 2015, September 24-25, Balatonfüred/Hungary

[6] I. FELEA, I. CSUZI, C. SECUI, E. BARLA: *Optimal dimensioning of public urban company tram parks*, "Journal of Sustainable Energy", ISSN 2067-5534, University of Oradea Editing House, 2015, Vol. VI, No.1, pp.6-11, Oradea / Romania

[7] ENESCO industrial: *Indicatori ODEX pentru eficiență energetică-RO;*(EN-*ODEX indicators for energy efficency)* www.enescoindustrial.com

[8] ******: *Sustainable economics with clean and energy efficient vehicles*, EU Commission, MEMO/07/594, Brussels/Belgium, 19th December 2007,

[9] ******: Key World Energy Statistic 2016, Edited by International Energy Agency - IEA, Free publications-statistics, 75739 Paris/France,Cedex 15,

[10] *****: *"3ibs – the intelligent, innovative, integrated Bus Systems"*, Research Program co-funded by the European Commission under 7th Research and

Technological Development Framework Program FP7, Research and Innovation Directorate-General, UITP papers, Brussels/Belgium, 2017

[11] A.MÜLLER-HELLMANN: *Towards zero/ low carbon urban mobility - Innovative multimodal electric urban public transport systems*, 60th UITP World Congress papers, May 2013, Geneva/ Switzerland

[12] *****: *"EBSF_2 – the European Bus System of Future 2"*, European Union's Horizon 2020 Research and Innovations Program under grant agreement No.63300, UITP papers, Brussels/Belgium, 2015

[13] *****: *"ELIPTIC - electrification of public transport in cities"*, European Union's Horizon 2020 Research and Innovations Program under grant agreement No.636012, UITP papers, Brussels, 2015

[14] *****: *"ZeEUS – Zero Emission Urban Bus Systems"*, European Union's Horizon 2020 Research and Innovations Program under grant agreement No.605485, UITP papers, Brussels/Belgium, 2016

[15] A.TOMASEVIC, S.MISANOVIC: *The first e-bus line in Belgrade, contribution to the sustainable development of the city*, "Chariot car" Ltd. presentation, Belgrad/Serbia, 2016

[16] L.LIBOR, I.CSUZI: *Oradea urban bus system energy efficiency analysis*, "Journal of Sustainable Energy", ISSN 2067-5534, University of Oradea Editing House, 2017, Vol. VIII, No.2, pp. 72-75, Oradea/ ROMANIA,

[17] D.COADY, I.PARRY, L.SEARS, B.SHANG: *How Large Are Global Energy Subsidies?* IMF Working Papers nr.15/105, Washington DC/ USA, May 2015

[18] *****: *A Roadmap for moving to a competitive low carbon economy in 2050*, EU Comission COM (2011), 112 final, 8.3.2011, Brussels/Belgium

AUTOBIOGRAPHY:

dr. eng. **István CSUZI**

WORK EXPERIENCES:

May 2017 –today: General Manager UAE (CITY TRANSPORT in ABU DHABI/United Arab Emirates - public transport company, as a member of Emirates National Group, https://enguae.com/board-of-directors)

Dec 2008 –Apr 2017: General Manager (OTL SA - Public Transport Company in ORADEA, Romania, www.otl.ro)

Oct 1992 –Dec 2008: General Manager (Management), TRIODA Ltd ORADEA, Romania, Electrical & Electronic components distributor (www.trioda.ro)

Sep 1988 –Oct 1992: Service Engineer (Maintenance & Management), Automatics & Industrial Power Electronics, IIRUC SA BUCHAREST, Romania (www.iiruc.ro) **Aug 1985 –Sep 1988**: Service Engineer (Maintenance & Management), Chief of Energetic at 3000 t clinker oven, CIMENT Factory ALESD, Romania (www.holcim.ro)

PROFESSIONAL RESULTS:

- Invention Patent Nr.47/1988, Bucharest, Romania (Adjusting the speed criteria for self-feeding bunker by the weight of the material in clinker production technology)

EDUCATION:

Sep 2012 –today: Associate Professor (Education), UNIVERSITY of ORADEA, Faculty of ENERGY and INDUSTRIAL MANAGEMENT, (www.uoradea.ro)

Sep 2007 –Dec 2011, PhD, Thesis title: "Contributions to evaluating and optimizing the energetic and availability performances of the urban electrical traction system", University of ORADEA, Romania (www.uoradea.ro)

Sep 1980 –Jul 1985, Diplomat Engineer in Electronics & Telecommunications, IPTV-Faculty of Electronics TIMISOARA, Romania (www.upt.ro)

MEMBERSHIPS:

Dec 2008 –May 2017: Member of Executive Board of URTP (Romanian Public Transport Union)

Feb 2009 –today: Member of Romanian General Association of Engineers

Oct 2010 –today: Member of Light Rail Train Committee of UITP (Union International de Transport Public)

June 2013 –May 2017: Chair of Light Rail Assembly UITP

June 2013 –May 2017: Vice President of UITP, member of Executive Board

Apr 2014 –May 2017: Vice President of URTP

March 2016 –today: President of Hungarian Science Society of Transylvania / Bihor county / Romania

March 2017 –today: Member in the Scientific Council of Doctoral School University of Oradea

May 2017 –today: external member of the Hungarian Scientific Academy in the Regional Committee for Hungarians Abroad, Budapest/HUNGARY

CONTACT emails:
gmuae@citytransport.ae , istvancsuzi@yahoo.com

TABLE OF CONTENTS: